Multiplication

Practice Workbook

on Graph Paper

- 1-Digit & 2-Digit Multipliers
- 3-Digit Multipliers in the Challenge Section
- Grid Lines for Easy Number Alignment
- Answer Key

TABLE OF CONTENTS

SECTION

1

Multiplication

2-Digits x 1-Digit

$$\begin{array}{r} {}^{3} \\ 76 \\ \times\ 5 \\ \hline 380 \end{array}$$

Multiply the 5 by each digit of 76 beginning with the digit in the ones position and carry as needed.

Multiply: 5 x 6 = 30

Write the 0 in the ones column and carry the 3 (above the 7).

Multiply: 5 x 7 = 35

Add the 3 that was carried in the previous step: 35 + 3 = 38

Write 38 in the next two columns.

Name ______________________ Date ______________________

Multiply.

① 32 × 4	② 91 × 3	③ 23 × 5	④ 93 × 7	⑤ 54 × 2
⑥ 82 × 3	⑦ 27 × 6	⑧ 22 × 4	⑨ 27 × 2	⑩ 83 × 4
⑪ 59 × 5	⑫ 98 × 2	⑬ 51 × 9	⑭ 18 × 8	⑮ 49 × 2
⑯ 49 × 7	⑰ 28 × 5	⑱ 15 × 9	⑲ 27 × 7	⑳ 69 × 7

Name ______________________ Date ______________

Multiply.

① 43 × 3

② 50 × 5

③ 29 × 2

④ 83 × 6

⑤ 54 × 5

⑥ 37 × 8

⑦ 46 × 7

⑧ 81 × 3

⑨ 47 × 4

⑩ 96 × 5

⑪ 17 × 6

⑫ 36 × 4

⑬ 54 × 9

⑭ 85 × 4

⑮ 67 × 6

⑯ 56 × 8

⑰ 97 × 3

⑱ 87 × 6

⑲ 14 × 5

⑳ 74 × 2

Name ____________________ Date ____________________

Multiply.

(1) 35 × 5	(2) 59 × 6	(3) 71 × 6	(4) 92 × 2	(5) 64 × 7
(6) 43 × 6	(7) 58 × 5	(8) 56 × 2	(9) 30 × 7	(10) 26 × 4
(11) 31 × 2	(12) 29 × 8	(13) 22 × 5	(14) 93 × 6	(15) 44 × 5
(16) 16 × 3	(17) 52 × 8	(18) 92 × 7	(19) 91 × 5	(20) 57 × 7

Name ____________________ Date ____________________

Multiply.

(1) 59×2

(2) 45×4

(3) 14×2

(4) 59×4

(5) 33×2

(6) 48×9

(7) 83×5

(8) 25×4

(9) 61×5

(10) 29×6

(11) 64×2

(12) 52×6

(13) 85×8

(14) 35×7

(15) 39×4

(16) 47×9

(17) 18×3

(18) 79×7

(19) 38×2

(20) 22×9

Name ______________________ Date ______________________

Multiply.

1) 78×2

2) 39×6

3) 88×9

4) 57×6

5) 21×4

6) 76×3

7) 30×5

8) 39×8

9) 92×8

10) 42×5

11) 58×3

12) 59×8

13) 51×2

14) 63×3

15) 52×9

16) 65×5

17) 76×4

18) 37×4

19) 88×5

20) 57×3

Name ______________________ Date ______________________

Multiply.

① 24×3

② 82×5

③ 64×8

④ 69×2

⑤ 25×5

⑥ 71×5

⑦ 64×4

⑧ 92×4

⑨ 62×9

⑩ 41×8

⑪ 82×2

⑫ 34×5

⑬ 27×9

⑭ 56×9

⑮ 54×7

⑯ 28×3

⑰ 59×9

⑱ 57×5

⑲ 24×6

⑳ 32×3

Name ______________________ Date ______________

Multiply.

① 62×8	② 72×5	③ 92×9	④ 56×3	⑤ 68×3
⑥ 79×8	⑦ 93×4	⑧ 48×7	⑨ 54×4	⑩ 60×5
⑪ 87×5	⑫ 63×2	⑬ 23×3	⑭ 94×3	⑮ 34×7
⑯ 76×9	⑰ 87×2	⑱ 95×4	⑲ 52×2	⑳ 48×5

Name ______________________ Date ______________

Multiply.

(1) 25 × 8	(2) 98 × 6	(3) 75 × 2	(4) 91 × 2	(5) 45 × 8
(6) 63 × 7	(7) 13 × 7	(8) 95 × 2	(9) 24 × 8	(10) 70 × 6
(11) 64 × 6	(12) 75 × 8	(13) 38 × 5	(14) 97 × 9	(15) 44 × 8
(16) 39 × 7	(17) 56 × 6	(18) 37 × 2	(19) 71 × 4	(20) 87 × 3

Name ______________________ Date ______________

Multiply.

① 78×6	② 81×7	③ 43×7	④ 33×8	⑤ 27×8
⑥ 53×2	⑦ 21×2	⑧ 73×9	⑨ 24×4	⑩ 13×2
⑪ 74×6	⑫ 95×8	⑬ 16×5	⑭ 68×9	⑮ 81×6
⑯ 29×5	⑰ 95×5	⑱ 63×5	⑲ 42×4	⑳ 73×2

Name ______________________ Date ______________________

Multiply.

① 57 × 8	② 77 × 9	③ 58 × 4	④ 25 × 2	⑤ 75 × 5
⑥ 16 × 9	⑦ 33 × 7	⑧ 84 × 7	⑨ 93 × 8	⑩ 55 × 3
⑪ 32 × 9	⑫ 89 × 4	⑬ 38 × 4	⑭ 44 × 3	⑮ 55 × 2
⑯ 22 × 7	⑰ 48 × 3	⑱ 34 × 4	⑲ 15 × 7	⑳ 18 × 9

Name ______________________ Date ______________

Multiply.

① 15×8	② 71×2	③ 40×4	④ 39×3	⑤ 26×2
⑥ 28×7	⑦ 23×2	⑧ 93×5	⑨ 59×3	⑩ 67×8
⑪ 87×8	⑫ 42×6	⑬ 74×8	⑭ 84×6	⑮ 19×2
⑯ 69×8	⑰ 21×5	⑱ 67×5	⑲ 36×2	⑳ 87×7

Name ____________________ Date ____________________

Multiply.

1. 62×4

2. 57×2

3. 55×7

4. 17×4

5. 24×5

6. 81×2

7. 64×9

8. 73×7

9. 52×7

10. 53×9

11. 76×7

12. 45×2

13. 62×2

14. 44×6

15. 67×2

16. 25×7

17. 15×4

18. 84×9

19. 47×8

20. 36×9

Name ______________________ Date ______________

Multiply.

① 52×5

② 31×6

③ 55×4

④ 48×2

⑤ 75×9

⑥ 17×7

⑦ 19×5

⑧ 43×4

⑨ 25×9

⑩ 65×2

⑪ 58×7

⑫ 21×8

⑬ 68×5

⑭ 35×9

⑮ 51×3

⑯ 18×4

⑰ 37×3

⑱ 25×3

⑲ 34×6

⑳ 40×8

Name ________________ Date ________________

Multiply.

① 85×9

② 82×4

③ 17×3

④ 73×6

⑤ 94×2

⑥ 53×5

⑦ 74×5

⑧ 83×8

⑨ 35×2

⑩ 84×5

⑪ 55×6

⑫ 74×3

⑬ 29×9

⑭ 72×4

⑮ 89×3

⑯ 79×4

⑰ 27×5

⑱ 81×4

⑲ 28×9

⑳ 83×2

Name ____________________ Date ____________________

Multiply.

(1) $\begin{array}{r} 75 \\ \times\ 3 \\ \hline \end{array}$

(2) $\begin{array}{r} 95 \\ \times\ 3 \\ \hline \end{array}$

(3) $\begin{array}{r} 28 \\ \times\ 2 \\ \hline \end{array}$

(4) $\begin{array}{r} 43 \\ \times\ 8 \\ \hline \end{array}$

(5) $\begin{array}{r} 99 \\ \times\ 4 \\ \hline \end{array}$

(6) $\begin{array}{r} 94 \\ \times\ 7 \\ \hline \end{array}$

(7) $\begin{array}{r} 76 \\ \times\ 8 \\ \hline \end{array}$

(8) $\begin{array}{r} 19 \\ \times\ 7 \\ \hline \end{array}$

(9) $\begin{array}{r} 31 \\ \times\ 3 \\ \hline \end{array}$

(10) $\begin{array}{r} 39 \\ \times\ 5 \\ \hline \end{array}$

(11) $\begin{array}{r} 31 \\ \times\ 7 \\ \hline \end{array}$

(12) $\begin{array}{r} 69 \\ \times\ 3 \\ \hline \end{array}$

(13) $\begin{array}{r} 97 \\ \times\ 6 \\ \hline \end{array}$

(14) $\begin{array}{r} 77 \\ \times\ 8 \\ \hline \end{array}$

(15) $\begin{array}{r} 44 \\ \times\ 9 \\ \hline \end{array}$

(16) $\begin{array}{r} 85 \\ \times\ 2 \\ \hline \end{array}$

(17) $\begin{array}{r} 43 \\ \times\ 2 \\ \hline \end{array}$

(18) $\begin{array}{r} 39 \\ \times\ 2 \\ \hline \end{array}$

(19) $\begin{array}{r} 65 \\ \times\ 7 \\ \hline \end{array}$

(20) $\begin{array}{r} 37 \\ \times\ 5 \\ \hline \end{array}$

Name ______________________ Date ______________________

Multiply.

① $\begin{array}{r} 28 \\ \times\ 4 \\ \hline \end{array}$

② $\begin{array}{r} 32 \\ \times\ 2 \\ \hline \end{array}$

③ $\begin{array}{r} 96 \\ \times\ 7 \\ \hline \end{array}$

④ $\begin{array}{r} 82 \\ \times\ 7 \\ \hline \end{array}$

⑤ $\begin{array}{r} 35 \\ \times\ 8 \\ \hline \end{array}$

⑥ $\begin{array}{r} 61 \\ \times\ 8 \\ \hline \end{array}$

⑦ $\begin{array}{r} 86 \\ \times\ 2 \\ \hline \end{array}$

⑧ $\begin{array}{r} 46 \\ \times\ 8 \\ \hline \end{array}$

⑨ $\begin{array}{r} 91 \\ \times\ 7 \\ \hline \end{array}$

⑩ $\begin{array}{r} 86 \\ \times\ 5 \\ \hline \end{array}$

⑪ $\begin{array}{r} 53 \\ \times\ 3 \\ \hline \end{array}$

⑫ $\begin{array}{r} 45 \\ \times\ 7 \\ \hline \end{array}$

⑬ $\begin{array}{r} 94 \\ \times\ 5 \\ \hline \end{array}$

⑭ $\begin{array}{r} 73 \\ \times\ 8 \\ \hline \end{array}$

⑮ $\begin{array}{r} 88 \\ \times\ 6 \\ \hline \end{array}$

⑯ $\begin{array}{r} 95 \\ \times\ 9 \\ \hline \end{array}$

⑰ $\begin{array}{r} 41 \\ \times\ 5 \\ \hline \end{array}$

⑱ $\begin{array}{r} 45 \\ \times\ 5 \\ \hline \end{array}$

⑲ $\begin{array}{r} 81 \\ \times\ 8 \\ \hline \end{array}$

⑳ $\begin{array}{r} 29 \\ \times\ 3 \\ \hline \end{array}$

Name ______________________ Date ______________________

Multiply.

① 23 × 8

② 55 × 9

③ 49 × 4

④ 17 × 5

⑤ 95 × 7

⑥ 82 × 6

⑦ 56 × 4

⑧ 31 × 8

⑨ 16 × 2

⑩ 75 × 7

⑪ 85 × 7

⑫ 36 × 5

⑬ 47 × 6

⑭ 94 × 8

⑮ 81 × 9

⑯ 55 × 8

⑰ 42 × 7

⑱ 21 × 3

⑲ 82 × 8

⑳ 45 × 6

Name ______________________ Date ______________________

Multiply.

(1) 58×2

(2) 62×3

(3) 92×3

(4) 89×9

(5) 62×5

(6) 15×2

(7) 43×5

(8) 54×6

(9) 38×6

(10) 89×8

(11) 70×3

(12) 96×9

(13) 61×2

(14) 27×4

(15) 51×7

(16) 86×8

(17) 14×7

(18) 24×2

(19) 26×7

(20) 83×9

Name ______________________ Date ______________

Multiply.

① 32 × 7	② 18 × 2	③ 48 × 4	④ 33 × 6	⑤ 46 × 2
⑥ 93 × 9	⑦ 25 × 6	⑧ 34 × 3	⑨ 36 × 7	⑩ 98 × 5
⑪ 54 × 3	⑫ 27 × 3	⑬ 81 × 5	⑭ 45 × 3	⑮ 62 × 7
⑯ 49 × 5	⑰ 83 × 3	⑱ 74 × 7	⑲ 32 × 6	⑳ 18 × 7

Name ______________________ Date ______________

Multiply.

① 75×6	② 72×9	③ 35×4	④ 61×7	⑤ 46×5
⑥ 17×8	⑦ 23×9	⑧ 68×8	⑨ 13×8	⑩ 97×4
⑪ 31×5	⑫ 89×7	⑬ 34×2	⑭ 15×3	⑮ 54×8
⑯ 74×4	⑰ 42×9	⑱ 58×9	⑲ 77×4	⑳ 19×3

Name ______________________ Date ______________________

Multiply.

(1) 86×9

(2) 87×4

(3) 58×8

(4) 57×9

(5) 65×4

(6) 14×3

(7) 53×8

(8) 26×9

(9) 51×5

(10) 96×8

(11) 77×5

(12) 89×2

(13) 94×4

(14) 49×8

(15) 29×4

(16) 35×3

(17) 85×3

(18) 98×7

(19) 26×3

(20) 73×5

SECTION

2 Multilication

3-Digits x 1-Digit
4-Digits x 1-Digit

	3	2	
	8	4	3
	x		7
5	9	0	1

Multiply the 7 by each digit of 843 beginning with the digit in the ones position and carry as needed.

Multiply: 7 x 3 = 21

Write the 1 in the ones column and carry the 2 (above the 4).

Multiply: 7 x 4 = 28

Add the 2 that was carried in the previous step: 28 + 2 = 30

Write the 0 in the tens column and carry the 3 (above the 8).

Multiply: 7 x 8 = 56

Add the 3 that was carried in the previous step: 56 + 3 = 59

Write 59 in the next two columns.

Name ______________________ Date ______________

Multiply.

(1) 292×5

(2) 861×9

(3) 868×3

(4) 479×8

(5) 945×2

(6) 825×6

(7) 727×5

(8) 937×2

(9) 474×5

(10) 876×2

(11) 385×6

(12) 818×2

(13) 632×6

(14) 451×8

(15) 306×4

(16) 237×8

Name ______________________ Date ______________

Multiply.

① 235 × 7 =

② 212 × 6 =

③ 575 × 7 =

④ 683 × 4 =

⑤ 652 × 9 =

⑥ 481 × 6 =

⑦ 519 × 2 =

⑧ 476 × 7 =

⑨ 924 × 5 =

⑩ 658 × 3 =

⑪ 253 × 6 =

⑫ 904 × 2 =

⑬ 959 × 4 =

⑭ 565 × 4 =

⑮ 845 × 2 =

⑯ 615 × 7 =

Name ______________________ Date ______________________

Multiply.

1) 325 × 6

2) 673 × 7

3) 768 × 6

4) 672 × 9

5) 414 × 6

6) 981 × 2

7) 786 × 4

8) 395 × 8

9) 365 × 3

10) 184 × 7

11) 526 × 5

12) 275 × 8

13) 172 × 3

14) 459 × 3

15) 373 × 8

16) 601 × 6

Name ______________________ Date ______________

Multiply.

① 582 × 6	② 375 × 9	③ 236 × 7	④ 879 × 4
⑤ 183 × 4	⑥ 272 × 7	⑦ 432 × 5	⑧ 782 × 3
⑨ 758 × 2	⑩ 142 × 9	⑪ 371 × 4	⑫ 581 × 7
⑬ 973 × 5	⑭ 623 × 4	⑮ 691 × 3	⑯ 867 × 9

Name ______________________ Date ______________

Multiply.

① 317×9	② 724×8	③ 256×4	④ 163×5
⑤ 946×7	⑥ 648×9	⑦ 529×8	⑧ 542×5
⑨ 679×3	⑩ 414×8	⑪ 304×7	⑫ 255×4
⑬ 525×6	⑭ 219×3	⑮ 434×7	⑯ 415×2

Name ____________________ Date ____________________

Multiply.

① 321×9

② 623×5

③ 464×8

④ 189×5

⑤ 728×5

⑥ 916×3

⑦ 286×7

⑧ 391×3

⑨ 421×3

⑩ 401×3

⑪ 542×3

⑫ 238×4

⑬ 935×6

⑭ 135×8

⑮ 743×7

⑯ 213×8

Name ______________________ Date ______________________

Multiply.

① 273×8

② 508×4

③ 596×9

④ 492×9

⑤ 369×4

⑥ 931×7

⑦ 367×8

⑧ 813×7

⑨ 167×5

⑩ 485×9

⑪ 769×2

⑫ 182×3

⑬ 968×7

⑭ 521×8

⑮ 463×5

⑯ 138×7

Name ______________________ Date ______________

Multiply.

① 235×4	② 285×7	③ 324×2	④ 826×8
⑤ 352×6	⑥ 583×4	⑦ 954×7	⑧ 158×2
⑨ 315×7	⑩ 793×8	⑪ 936×2	⑫ 712×7
⑬ 698×2	⑭ 537×3	⑮ 236×9	⑯ 163×4

Name ______________________ Date ______________

Multiply.

① 961 × 2

② 297 × 8

③ 165 × 4

④ 542 × 9

⑤ 851 × 7

⑥ 579 × 4

⑦ 278 × 6

⑧ 818 × 5

⑨ 324 × 6

⑩ 637 × 4

⑪ 362 × 7

⑫ 605 × 6

⑬ 651 × 9

⑭ 543 × 8

⑮ 276 × 3

⑯ 396 × 3

Name ______________________ Date ______________

Multiply.

① $\begin{array}{r} 514 \\ \times\ \ 3 \\ \hline \end{array}$	② $\begin{array}{r} 687 \\ \times\ \ 2 \\ \hline \end{array}$	③ $\begin{array}{r} 428 \\ \times\ \ 4 \\ \hline \end{array}$	④ $\begin{array}{r} 267 \\ \times\ \ 3 \\ \hline \end{array}$
⑤ $\begin{array}{r} 248 \\ \times\ \ 2 \\ \hline \end{array}$	⑥ $\begin{array}{r} 875 \\ \times\ \ 8 \\ \hline \end{array}$	⑦ $\begin{array}{r} 494 \\ \times\ \ 9 \\ \hline \end{array}$	⑧ $\begin{array}{r} 462 \\ \times\ \ 4 \\ \hline \end{array}$
⑨ $\begin{array}{r} 635 \\ \times\ \ 7 \\ \hline \end{array}$	⑩ $\begin{array}{r} 485 \\ \times\ \ 3 \\ \hline \end{array}$	⑪ $\begin{array}{r} 761 \\ \times\ \ 2 \\ \hline \end{array}$	⑫ $\begin{array}{r} 574 \\ \times\ \ 8 \\ \hline \end{array}$
⑬ $\begin{array}{r} 489 \\ \times\ \ 2 \\ \hline \end{array}$	⑭ $\begin{array}{r} 482 \\ \times\ \ 9 \\ \hline \end{array}$	⑮ $\begin{array}{r} 521 \\ \times\ \ 5 \\ \hline \end{array}$	⑯ $\begin{array}{r} 868 \\ \times\ \ 7 \\ \hline \end{array}$

Name ______________________ Date ______________________

Multiply.

1) $\begin{array}{r} 352 \\ \times \quad 3 \\ \hline \end{array}$

2) $\begin{array}{r} 162 \\ \times \quad 8 \\ \hline \end{array}$

3) $\begin{array}{r} 274 \\ \times \quad 3 \\ \hline \end{array}$

4) $\begin{array}{r} 242 \\ \times \quad 5 \\ \hline \end{array}$

5) $\begin{array}{r} 808 \\ \times \quad 4 \\ \hline \end{array}$

6) $\begin{array}{r} 478 \\ \times \quad 6 \\ \hline \end{array}$

7) $\begin{array}{r} 921 \\ \times \quad 4 \\ \hline \end{array}$

8) $\begin{array}{r} 762 \\ \times \quad 5 \\ \hline \end{array}$

9) $\begin{array}{r} 521 \\ \times \quad 6 \\ \hline \end{array}$

10) $\begin{array}{r} 279 \\ \times \quad 5 \\ \hline \end{array}$

11) $\begin{array}{r} 574 \\ \times \quad 7 \\ \hline \end{array}$

12) $\begin{array}{r} 597 \\ \times \quad 8 \\ \hline \end{array}$

13) $\begin{array}{r} 362 \\ \times \quad 6 \\ \hline \end{array}$

14) $\begin{array}{r} 384 \\ \times \quad 5 \\ \hline \end{array}$

15) $\begin{array}{r} 126 \\ \times \quad 5 \\ \hline \end{array}$

16) $\begin{array}{r} 508 \\ \times \quad 6 \\ \hline \end{array}$

Name ______________________ Date ______________

Multiply.

① 3151 × 8 =

② 2508 × 3 =

③ 6749 × 2 =

④ 4763 × 7 =

⑤ 8437 × 8 =

⑥ 7532 × 3 =

⑦ 8651 × 5 =

⑧ 6737 × 9 =

⑨ 5059 × 7 =

⑩ 3932 × 8 =

⑪ 7817 × 2 =

⑫ 3186 × 4 =

Name ______________________ Date ______________________

Multiply.

① 6725 × 8

② 3841 × 9

③ 5195 × 7

④ 6848 × 5

⑤ 9682 × 6

⑥ 1943 × 4

⑦ 8589 × 3

⑧ 3783 × 2

⑨ 4704 × 3

⑩ 7567 × 4

⑪ 3712 × 8

⑫ 8358 × 2

Name ____________________ Date ____________

Multiply.

① 8949 × 6 =

② 1271 × 7 =

③ 4543 × 4 =

④ 5232 × 5 =

⑤ 2692 × 6 =

⑥ 9072 × 3 =

⑦ 8051 × 8 =

⑧ 3568 × 2 =

⑨ 6439 × 8 =

⑩ 1546 × 3 =

⑪ 7457 × 8 =

⑫ 7248 × 5 =

Name ____________________ Date ____________________

Multiply.

1) 4949 × 6

2) 3517 × 4

3) 4743 × 9

4) 6573 × 5

5) 3691 × 5

6) 7408 × 6

7) 9875 × 7

8) 8582 × 3

9) 9837 × 2

10) 5987 × 3

11) 8487 × 8

12) 6489 × 3

Name ______________________ Date ______________

Multiply.

① 3076×9	② 6792×7	③ 5681×3
④ 7526×8	⑤ 4743×5	⑥ 9795×4
⑦ 6341×9	⑧ 4432×6	⑨ 5738×9
⑩ 1671×2	⑪ 5417×3	⑫ 9242×3

Name ______________________ Date ______________

Multiply.

① 4962 × 6

② 5475 × 3

③ 5015 × 9

④ 4793 × 9

⑤ 3465 × 4

⑥ 4634 × 7

⑦ 2626 × 4

⑧ 9826 × 3

⑨ 3273 × 2

⑩ 8936 × 2

⑪ 7271 × 2

⑫ 6213 × 7

Name ______________________ Date ______________

Multiply.

1.
$$\begin{array}{r} 2545 \\ \times \quad 6 \\ \hline \end{array}$$

2.
$$\begin{array}{r} 7631 \\ \times \quad 4 \\ \hline \end{array}$$

3.
$$\begin{array}{r} 2744 \\ \times \quad 2 \\ \hline \end{array}$$

4.
$$\begin{array}{r} 8053 \\ \times \quad 8 \\ \hline \end{array}$$

5.
$$\begin{array}{r} 6946 \\ \times \quad 2 \\ \hline \end{array}$$

6.
$$\begin{array}{r} 1527 \\ \times \quad 6 \\ \hline \end{array}$$

7.
$$\begin{array}{r} 9583 \\ \times \quad 7 \\ \hline \end{array}$$

8.
$$\begin{array}{r} 1781 \\ \times \quad 6 \\ \hline \end{array}$$

9.
$$\begin{array}{r} 8364 \\ \times \quad 3 \\ \hline \end{array}$$

10.
$$\begin{array}{r} 4309 \\ \times \quad 4 \\ \hline \end{array}$$

11.
$$\begin{array}{r} 9381 \\ \times \quad 8 \\ \hline \end{array}$$

12.
$$\begin{array}{r} 4737 \\ \times \quad 4 \\ \hline \end{array}$$

Name ______________________ Date ______________

Multiply.

(1)
```
  5415
×    7
------
```

(2)
```
  6832
×    7
------
```

(3)
```
  2971
×    4
------
```

(4)
```
  6585
×    3
------
```

(5)
```
  1481
×    8
------
```

(6)
```
  5174
×    7
------
```

(7)
```
  7245
×    7
------
```

(8)
```
  4692
×    3
------
```

(9)
```
  3655
×    5
------
```

(10)
```
  4285
×    6
------
```

(11)
```
  7589
×    5
------
```

(12)
```
  2714
×    6
------
```

Name ______________________ Date ______________________

Multiply.

1)
```
  7235
×    2
------
```

2)
```
  4919
×    4
------
```

3)
```
  8187
×    2
------
```

4)
```
  1358
×    3
------
```

5)
```
  8654
×    7
------
```

6)
```
  7084
×    9
------
```

7)
```
  2809
×    2
------
```

8)
```
  2496
×    7
------
```

9)
```
  4762
×    5
------
```

10)
```
  5092
×    8
------
```

11)
```
  8714
×    3
------
```

12)
```
  1965
×    2
------
```

Name ______________________ Date ______________

Multiply.

① $\begin{array}{r} 6265 \\ \times \quad 4 \\ \hline \end{array}$	② $\begin{array}{r} 5874 \\ \times \quad 7 \\ \hline \end{array}$	③ $\begin{array}{r} 5277 \\ \times \quad 3 \\ \hline \end{array}$
④ $\begin{array}{r} 6519 \\ \times \quad 2 \\ \hline \end{array}$	⑤ $\begin{array}{r} 8185 \\ \times \quad 7 \\ \hline \end{array}$	⑥ $\begin{array}{r} 9463 \\ \times \quad 6 \\ \hline \end{array}$
⑦ $\begin{array}{r} 5804 \\ \times \quad 8 \\ \hline \end{array}$	⑧ $\begin{array}{r} 5768 \\ \times \quad 5 \\ \hline \end{array}$	⑨ $\begin{array}{r} 2956 \\ \times \quad 5 \\ \hline \end{array}$
⑩ $\begin{array}{r} 6917 \\ \times \quad 6 \\ \hline \end{array}$	⑪ $\begin{array}{r} 9168 \\ \times \quad 9 \\ \hline \end{array}$	⑫ $\begin{array}{r} 5876 \\ \times \quad 2 \\ \hline \end{array}$

SECTION

3

Multiplication

2-Digits x 2-Digits

		1	
		9	2
	x	4	8
	7	3	6
3	6	8	0
4,	4	1	6

STEP 1

Multiply by the ones digit of 48. *(See section 1 for instructions on multiplying 2-digit numbers by 1-digit numbers.)*

Multiply: 8 x 92 = 736

STEP 2

Multiply by the tens digit of 48. Write the product on the next row with 0 in the ones column (since 4 is the tens digit).

Write 0 in the ones column on the next row.
Multiply: 4 x 92 = 368 *(We have 3,680 since there is a 0 in the ones place.)*

STEP 3

Add the two rows.

Add: 736 + 3,680 = 4,416

Name ____________________ Date ____________________

Multiply.

(1) 72×76

(2) 48×45

(3) 19×16

(4) 85×37

(5) 13×68

(6) 45×63

(7) 18×54

(8) 17×48

(9) 51×67

(10) 48×28

(11) 75×39

(12) 47×59

(13) 92×32

(14) 83×34

(15) 75×37

(16) 24×24

Name ____________________ Date ____________________

Multiply.

(1) 94×31

(2) 53×53

(3) 91×43

(4) 37×56

(5) 74×62

(6) 41×53

(7) 46×36

(8) 64×93

(9) 59×81

(10) 57×24

(11) 41×97

(12) 95×38

(13) 63×94

(14) 13×27

(15) 87×34

(16) 45×83

Name ____________________ Date ____________________

Multiply.

(1)
$$\begin{array}{r} 26 \\ \times\ 72 \\ \hline \end{array}$$

(2)
$$\begin{array}{r} 13 \\ \times\ 56 \\ \hline \end{array}$$

(3)
$$\begin{array}{r} 73 \\ \times\ 62 \\ \hline \end{array}$$

(4)
$$\begin{array}{r} 37 \\ \times\ 87 \\ \hline \end{array}$$

(5)
$$\begin{array}{r} 26 \\ \times\ 27 \\ \hline \end{array}$$

(6)
$$\begin{array}{r} 69 \\ \times\ 46 \\ \hline \end{array}$$

(7)
$$\begin{array}{r} 51 \\ \times\ 85 \\ \hline \end{array}$$

(8)
$$\begin{array}{r} 48 \\ \times\ 89 \\ \hline \end{array}$$

(9)
$$\begin{array}{r} 47 \\ \times\ 91 \\ \hline \end{array}$$

(10)
$$\begin{array}{r} 57 \\ \times\ 42 \\ \hline \end{array}$$

(11)
$$\begin{array}{r} 32 \\ \times\ 35 \\ \hline \end{array}$$

(12)
$$\begin{array}{r} 32 \\ \times\ 72 \\ \hline \end{array}$$

(13)
$$\begin{array}{r} 35 \\ \times\ 52 \\ \hline \end{array}$$

(14)
$$\begin{array}{r} 29 \\ \times\ 25 \\ \hline \end{array}$$

(15)
$$\begin{array}{r} 72 \\ \times\ 81 \\ \hline \end{array}$$

(16)
$$\begin{array}{r} 72 \\ \times\ 26 \\ \hline \end{array}$$

Name ________________ Date ________________

Multiply.

(1) $\begin{array}{r} 27 \\ \times 23 \\ \hline \end{array}$	(2) $\begin{array}{r} 89 \\ \times 75 \\ \hline \end{array}$	(3) $\begin{array}{r} 46 \\ \times 24 \\ \hline \end{array}$	(4) $\begin{array}{r} 31 \\ \times 52 \\ \hline \end{array}$
(5) $\begin{array}{r} 27 \\ \times 56 \\ \hline \end{array}$	(6) $\begin{array}{r} 89 \\ \times 23 \\ \hline \end{array}$	(7) $\begin{array}{r} 42 \\ \times 39 \\ \hline \end{array}$	(8) $\begin{array}{r} 32 \\ \times 89 \\ \hline \end{array}$
(9) $\begin{array}{r} 24 \\ \times 67 \\ \hline \end{array}$	(10) $\begin{array}{r} 37 \\ \times 83 \\ \hline \end{array}$	(11) $\begin{array}{r} 62 \\ \times 13 \\ \hline \end{array}$	(12) $\begin{array}{r} 39 \\ \times 47 \\ \hline \end{array}$
(13) $\begin{array}{r} 83 \\ \times 67 \\ \hline \end{array}$	(14) $\begin{array}{r} 35 \\ \times 79 \\ \hline \end{array}$	(15) $\begin{array}{r} 57 \\ \times 37 \\ \hline \end{array}$	(16) $\begin{array}{r} 97 \\ \times 79 \\ \hline \end{array}$

Name ______________________ Date ______________________

Multiply.

1. 87×29

2. 52×59

3. 23×75

4. 91×76

5. 26×43

6. 89×48

7. 72×94

8. 51×29

9. 26×38

10. 25×52

11. 76×81

12. 18×78

13. 38×25

14. 65×43

15. 25×13

16. 53×56

Name ____________________ Date ____________________

Multiply.

(1) 56×28

(2) 46×94

(3) 85×34

(4) 46×98

(5) 72×27

(6) 82×15

(7) 41×59

(8) 52×96

(9) 34×57

(10) 79×45

(11) 24×68

(12) 87×37

(13) 25×54

(14) 58×65

(15) 61×27

(16) 79×25

Name ______________________ Date ______________

Multiply.

① $\begin{array}{r} 81 \\ \times\ 58 \\ \hline \end{array}$	② $\begin{array}{r} 16 \\ \times\ 48 \\ \hline \end{array}$	③ $\begin{array}{r} 35 \\ \times\ 59 \\ \hline \end{array}$	④ $\begin{array}{r} 26 \\ \times\ 31 \\ \hline \end{array}$
⑤ $\begin{array}{r} 49 \\ \times\ 39 \\ \hline \end{array}$	⑥ $\begin{array}{r} 59 \\ \times\ 82 \\ \hline \end{array}$	⑦ $\begin{array}{r} 75 \\ \times\ 25 \\ \hline \end{array}$	⑧ $\begin{array}{r} 62 \\ \times\ 65 \\ \hline \end{array}$
⑨ $\begin{array}{r} 53 \\ \times\ 42 \\ \hline \end{array}$	⑩ $\begin{array}{r} 14 \\ \times\ 16 \\ \hline \end{array}$	⑪ $\begin{array}{r} 86 \\ \times\ 37 \\ \hline \end{array}$	⑫ $\begin{array}{r} 25 \\ \times\ 94 \\ \hline \end{array}$
⑬ $\begin{array}{r} 18 \\ \times\ 45 \\ \hline \end{array}$	⑭ $\begin{array}{r} 19 \\ \times\ 63 \\ \hline \end{array}$	⑮ $\begin{array}{r} 28 \\ \times\ 86 \\ \hline \end{array}$	⑯ $\begin{array}{r} 93 \\ \times\ 62 \\ \hline \end{array}$

Name ____________________ Date ____________________

Multiply.

(1) 21×26

(2) 85×57

(3) 35×35

(4) 94×46

(5) 92×67

(6) 46×82

(7) 47×37

(8) 24×32

(9) 69×93

(10) 83×48

(11) 48×87

(12) 36×25

(13) 32×15

(14) 42×37

(15) 75×86

(16) 62×72

Name ______________________ Date ______________

Multiply.

(1) 47×84

(2) 53×46

(3) 75×83

(4) 91×83

(5) 84×47

(6) 51×27

(7) 47×17

(8) 29×69

(9) 74×64

(10) 93×65

(11) 15×48

(12) 74×23

(13) 42×71

(14) 86×63

(15) 43×74

(16) 47×72

Name ______________________ Date ______________

Multiply.

(1)
$$\begin{array}{r} 61 \\ \times\ 78 \\ \hline \end{array}$$

(2)
$$\begin{array}{r} 43 \\ \times\ 43 \\ \hline \end{array}$$

(3)
$$\begin{array}{r} 34 \\ \times\ 71 \\ \hline \end{array}$$

(4)
$$\begin{array}{r} 91 \\ \times\ 59 \\ \hline \end{array}$$

(5)
$$\begin{array}{r} 46 \\ \times\ 28 \\ \hline \end{array}$$

(6)
$$\begin{array}{r} 54 \\ \times\ 42 \\ \hline \end{array}$$

(7)
$$\begin{array}{r} 35 \\ \times\ 37 \\ \hline \end{array}$$

(8)
$$\begin{array}{r} 31 \\ \times\ 75 \\ \hline \end{array}$$

(9)
$$\begin{array}{r} 25 \\ \times\ 51 \\ \hline \end{array}$$

(10)
$$\begin{array}{r} 83 \\ \times\ 68 \\ \hline \end{array}$$

(11)
$$\begin{array}{r} 83 \\ \times\ 78 \\ \hline \end{array}$$

(12)
$$\begin{array}{r} 52 \\ \times\ 48 \\ \hline \end{array}$$

(13)
$$\begin{array}{r} 16 \\ \times\ 76 \\ \hline \end{array}$$

(14)
$$\begin{array}{r} 56 \\ \times\ 45 \\ \hline \end{array}$$

(15)
$$\begin{array}{r} 71 \\ \times\ 23 \\ \hline \end{array}$$

(16)
$$\begin{array}{r} 92 \\ \times\ 35 \\ \hline \end{array}$$

Name ______________________ Date ______________

Multiply.

(1) $\begin{array}{r} 31 \\ \times\ 64 \\ \hline \end{array}$	(2) $\begin{array}{r} 45 \\ \times\ 42 \\ \hline \end{array}$	(3) $\begin{array}{r} 31 \\ \times\ 67 \\ \hline \end{array}$	(4) $\begin{array}{r} 57 \\ \times\ 85 \\ \hline \end{array}$
(5) $\begin{array}{r} 17 \\ \times\ 23 \\ \hline \end{array}$	(6) $\begin{array}{r} 29 \\ \times\ 81 \\ \hline \end{array}$	(7) $\begin{array}{r} 82 \\ \times\ 92 \\ \hline \end{array}$	(8) $\begin{array}{r} 28 \\ \times\ 96 \\ \hline \end{array}$
(9) $\begin{array}{r} 58 \\ \times\ 49 \\ \hline \end{array}$	(10) $\begin{array}{r} 36 \\ \times\ 29 \\ \hline \end{array}$	(11) $\begin{array}{r} 71 \\ \times\ 71 \\ \hline \end{array}$	(12) $\begin{array}{r} 68 \\ \times\ 43 \\ \hline \end{array}$
(13) $\begin{array}{r} 94 \\ \times\ 69 \\ \hline \end{array}$	(14) $\begin{array}{r} 46 \\ \times\ 57 \\ \hline \end{array}$	(15) $\begin{array}{r} 27 \\ \times\ 28 \\ \hline \end{array}$	(16) $\begin{array}{r} 56 \\ \times\ 32 \\ \hline \end{array}$

Name ______________________ Date ______________

Multiply.

① 34×78

② 61×35

③ 23×84

④ 59×31

⑤ 93×54

⑥ 56×54

⑦ 14×42

⑧ 72×73

⑨ 53×37

⑩ 81×56

⑪ 36×54

⑫ 34×75

⑬ 62×47

⑭ 83×97

⑮ 46×21

⑯ 93×47

Name ______________________ Date ______________

Multiply.

1. $\begin{array}{r} 14 \\ \times\ 23 \\ \hline \end{array}$

2. $\begin{array}{r} 83 \\ \times\ 95 \\ \hline \end{array}$

3. $\begin{array}{r} 42 \\ \times\ 94 \\ \hline \end{array}$

4. $\begin{array}{r} 72 \\ \times\ 85 \\ \hline \end{array}$

5. $\begin{array}{r} 47 \\ \times\ 82 \\ \hline \end{array}$

6. $\begin{array}{r} 93 \\ \times\ 53 \\ \hline \end{array}$

7. $\begin{array}{r} 58 \\ \times\ 79 \\ \hline \end{array}$

8. $\begin{array}{r} 15 \\ \times\ 25 \\ \hline \end{array}$

9. $\begin{array}{r} 31 \\ \times\ 13 \\ \hline \end{array}$

10. $\begin{array}{r} 78 \\ \times\ 43 \\ \hline \end{array}$

11. $\begin{array}{r} 96 \\ \times\ 74 \\ \hline \end{array}$

12. $\begin{array}{r} 27 \\ \times\ 49 \\ \hline \end{array}$

13. $\begin{array}{r} 24 \\ \times\ 15 \\ \hline \end{array}$

14. $\begin{array}{r} 35 \\ \times\ 92 \\ \hline \end{array}$

15. $\begin{array}{r} 74 \\ \times\ 43 \\ \hline \end{array}$

16. $\begin{array}{r} 94 \\ \times\ 37 \\ \hline \end{array}$

Name ______________________ Date ______________

Multiply.

① $\begin{array}{r} 26 \\ \times\ 69 \\ \hline \end{array}$

② $\begin{array}{r} 41 \\ \times\ 46 \\ \hline \end{array}$

③ $\begin{array}{r} 35 \\ \times\ 54 \\ \hline \end{array}$

④ $\begin{array}{r} 81 \\ \times\ 54 \\ \hline \end{array}$

⑤ $\begin{array}{r} 37 \\ \times\ 59 \\ \hline \end{array}$

⑥ $\begin{array}{r} 28 \\ \times\ 92 \\ \hline \end{array}$

⑦ $\begin{array}{r} 95 \\ \times\ 62 \\ \hline \end{array}$

⑧ $\begin{array}{r} 42 \\ \times\ 51 \\ \hline \end{array}$

⑨ $\begin{array}{r} 87 \\ \times\ 73 \\ \hline \end{array}$

⑩ $\begin{array}{r} 59 \\ \times\ 58 \\ \hline \end{array}$

⑪ $\begin{array}{r} 28 \\ \times\ 53 \\ \hline \end{array}$

⑫ $\begin{array}{r} 73 \\ \times\ 46 \\ \hline \end{array}$

⑬ $\begin{array}{r} 21 \\ \times\ 73 \\ \hline \end{array}$

⑭ $\begin{array}{r} 86 \\ \times\ 32 \\ \hline \end{array}$

⑮ $\begin{array}{r} 79 \\ \times\ 41 \\ \hline \end{array}$

⑯ $\begin{array}{r} 46 \\ \times\ 52 \\ \hline \end{array}$

Name ______________________ Date ______________

Multiply.

(1) $\begin{array}{r} 42 \\ \times 78 \\ \hline \end{array}$

(2) $\begin{array}{r} 96 \\ \times 45 \\ \hline \end{array}$

(3) $\begin{array}{r} 25 \\ \times 87 \\ \hline \end{array}$

(4) $\begin{array}{r} 14 \\ \times 14 \\ \hline \end{array}$

(5) $\begin{array}{r} 23 \\ \times 35 \\ \hline \end{array}$

(6) $\begin{array}{r} 91 \\ \times 87 \\ \hline \end{array}$

(7) $\begin{array}{r} 93 \\ \times 79 \\ \hline \end{array}$

(8) $\begin{array}{r} 41 \\ \times 23 \\ \hline \end{array}$

(9) $\begin{array}{r} 84 \\ \times 36 \\ \hline \end{array}$

(10) $\begin{array}{r} 25 \\ \times 45 \\ \hline \end{array}$

(11) $\begin{array}{r} 86 \\ \times 57 \\ \hline \end{array}$

(12) $\begin{array}{r} 54 \\ \times 56 \\ \hline \end{array}$

(13) $\begin{array}{r} 82 \\ \times 14 \\ \hline \end{array}$

(14) $\begin{array}{r} 41 \\ \times 93 \\ \hline \end{array}$

(15) $\begin{array}{r} 24 \\ \times 92 \\ \hline \end{array}$

(16) $\begin{array}{r} 95 \\ \times 56 \\ \hline \end{array}$

Name ______________________ Date ______________

Multiply.

(1) $\begin{array}{r} 39 \\ \times\ 68 \\ \hline \end{array}$

(2) $\begin{array}{r} 24 \\ \times\ 45 \\ \hline \end{array}$

(3) $\begin{array}{r} 74 \\ \times\ 51 \\ \hline \end{array}$

(4) $\begin{array}{r} 89 \\ \times\ 74 \\ \hline \end{array}$

(5) $\begin{array}{r} 79 \\ \times\ 74 \\ \hline \end{array}$

(6) $\begin{array}{r} 65 \\ \times\ 24 \\ \hline \end{array}$

(7) $\begin{array}{r} 39 \\ \times\ 46 \\ \hline \end{array}$

(8) $\begin{array}{r} 23 \\ \times\ 65 \\ \hline \end{array}$

(9) $\begin{array}{r} 54 \\ \times\ 21 \\ \hline \end{array}$

(10) $\begin{array}{r} 25 \\ \times\ 68 \\ \hline \end{array}$

(11) $\begin{array}{r} 14 \\ \times\ 78 \\ \hline \end{array}$

(12) $\begin{array}{r} 17 \\ \times\ 85 \\ \hline \end{array}$

(13) $\begin{array}{r} 67 \\ \times\ 49 \\ \hline \end{array}$

(14) $\begin{array}{r} 59 \\ \times\ 26 \\ \hline \end{array}$

(15) $\begin{array}{r} 83 \\ \times\ 80 \\ \hline \end{array}$

(16) $\begin{array}{r} 62 \\ \times\ 75 \\ \hline \end{array}$

Name ______________________ Date ______________

Multiply.

(1) 67×58

(2) 17×82

(3) 25×43

(4) 43×17

(5) 26×48

(6) 61×85

(7) 69×43

(8) 85×93

(9) 46×67

(10) 89×47

(11) 13×23

(12) 73×87

(13) 57×51

(14) 68×48

(15) 42×27

(16) 18×34

Name ____________________ Date ____________________

Multiply.

(1) $\begin{array}{r} 45 \\ \times 34 \\ \hline \end{array}$

(2) $\begin{array}{r} 46 \\ \times 15 \\ \hline \end{array}$

(3) $\begin{array}{r} 62 \\ \times 89 \\ \hline \end{array}$

(4) $\begin{array}{r} 51 \\ \times 68 \\ \hline \end{array}$

(5) $\begin{array}{r} 42 \\ \times 34 \\ \hline \end{array}$

(6) $\begin{array}{r} 25 \\ \times 64 \\ \hline \end{array}$

(7) $\begin{array}{r} 86 \\ \times 14 \\ \hline \end{array}$

(8) $\begin{array}{r} 74 \\ \times 52 \\ \hline \end{array}$

(9) $\begin{array}{r} 39 \\ \times 89 \\ \hline \end{array}$

(10) $\begin{array}{r} 25 \\ \times 91 \\ \hline \end{array}$

(11) $\begin{array}{r} 48 \\ \times 74 \\ \hline \end{array}$

(12) $\begin{array}{r} 27 \\ \times 63 \\ \hline \end{array}$

(13) $\begin{array}{r} 47 \\ \times 36 \\ \hline \end{array}$

(14) $\begin{array}{r} 78 \\ \times 39 \\ \hline \end{array}$

(15) $\begin{array}{r} 24 \\ \times 72 \\ \hline \end{array}$

(16) $\begin{array}{r} 35 \\ \times 75 \\ \hline \end{array}$

Name ____________________ Date ____________________

Multiply.

(1)
$$\begin{array}{r} 73 \\ \times\ 72 \\ \hline \end{array}$$

(2)
$$\begin{array}{r} 62 \\ \times\ 35 \\ \hline \end{array}$$

(3)
$$\begin{array}{r} 14 \\ \times\ 35 \\ \hline \end{array}$$

(4)
$$\begin{array}{r} 45 \\ \times\ 27 \\ \hline \end{array}$$

(5)
$$\begin{array}{r} 75 \\ \times\ 24 \\ \hline \end{array}$$

(6)
$$\begin{array}{r} 48 \\ \times\ 94 \\ \hline \end{array}$$

(7)
$$\begin{array}{r} 84 \\ \times\ 25 \\ \hline \end{array}$$

(8)
$$\begin{array}{r} 92 \\ \times\ 38 \\ \hline \end{array}$$

(9)
$$\begin{array}{r} 82 \\ \times\ 62 \\ \hline \end{array}$$

(10)
$$\begin{array}{r} 93 \\ \times\ 59 \\ \hline \end{array}$$

(11)
$$\begin{array}{r} 31 \\ \times\ 34 \\ \hline \end{array}$$

(12)
$$\begin{array}{r} 98 \\ \times\ 78 \\ \hline \end{array}$$

(13)
$$\begin{array}{r} 83 \\ \times\ 15 \\ \hline \end{array}$$

(14)
$$\begin{array}{r} 16 \\ \times\ 95 \\ \hline \end{array}$$

(15)
$$\begin{array}{r} 63 \\ \times\ 42 \\ \hline \end{array}$$

(16)
$$\begin{array}{r} 73 \\ \times\ 68 \\ \hline \end{array}$$

Name ______________________ Date ______________

Multiply.

① 38 × 63 = ____

② 56 × 24 = ____

③ 48 × 67 = ____

④ 87 × 36 = ____

⑤ 35 × 44 = ____

⑥ 41 × 75 = ____

⑦ 86 × 59 = ____

⑧ 52 × 47 = ____

⑨ 14 × 43 = ____

⑩ 61 × 46 = ____

⑪ 29 × 53 = ____

⑫ 37 × 46 = ____

⑬ 14 × 19 = ____

⑭ 56 × 97 = ____

⑮ 38 × 56 = ____

⑯ 68 × 35 = ____

Name ______________________ Date ______________

Multiply.

(1) $\begin{array}{r} 52 \\ \times\ 56 \\ \hline \end{array}$

(2) $\begin{array}{r} 38 \\ \times\ 84 \\ \hline \end{array}$

(3) $\begin{array}{r} 98 \\ \times\ 67 \\ \hline \end{array}$

(4) $\begin{array}{r} 53 \\ \times\ 29 \\ \hline \end{array}$

(5) $\begin{array}{r} 76 \\ \times\ 46 \\ \hline \end{array}$

(6) $\begin{array}{r} 24 \\ \times\ 25 \\ \hline \end{array}$

(7) $\begin{array}{r} 83 \\ \times\ 19 \\ \hline \end{array}$

(8) $\begin{array}{r} 82 \\ \times\ 93 \\ \hline \end{array}$

(9) $\begin{array}{r} 93 \\ \times\ 46 \\ \hline \end{array}$

(10) $\begin{array}{r} 37 \\ \times\ 24 \\ \hline \end{array}$

(11) $\begin{array}{r} 15 \\ \times\ 43 \\ \hline \end{array}$

(12) $\begin{array}{r} 31 \\ \times\ 35 \\ \hline \end{array}$

(13) $\begin{array}{r} 31 \\ \times\ 45 \\ \hline \end{array}$

(14) $\begin{array}{r} 86 \\ \times\ 39 \\ \hline \end{array}$

(15) $\begin{array}{r} 78 \\ \times\ 36 \\ \hline \end{array}$

(16) $\begin{array}{r} 52 \\ \times\ 25 \\ \hline \end{array}$

SECTION

4

Multiplication
3-Digits x 2-Digits

		4	5	
		1	1	
		1	5	8
		x	7	2
		3	1	6
1	1	0	6	0
1	1,	3	7	6

STEP 1
Multiply by the ones digit of 72.

Multiply: 2 x 158 = 316

STEP 2
Multiply by the tens digit of 72. Write the product on the next row with 0 in the ones column (since 7 is the tens digit).

Write 0 in the ones column on the next row.
Multiply: 7 x 158 = 1,106 *(We have 11,060 since 0 is in the ones place.)*

STEP 3
Add the two rows.

Add: 316 + 11,060 = 11,376

Name ______________________ Date ______________

Multiply.

(1)
$$\begin{array}{r} 191 \\ \times\ 34 \\ \hline \end{array}$$

(2)
$$\begin{array}{r} 639 \\ \times\ 37 \\ \hline \end{array}$$

(3)
$$\begin{array}{r} 203 \\ \times\ 84 \\ \hline \end{array}$$

(4)
$$\begin{array}{r} 358 \\ \times\ 46 \\ \hline \end{array}$$

(5)
$$\begin{array}{r} 734 \\ \times\ 38 \\ \hline \end{array}$$

(6)
$$\begin{array}{r} 938 \\ \times\ 92 \\ \hline \end{array}$$

(7)
$$\begin{array}{r} 531 \\ \times\ 75 \\ \hline \end{array}$$

(8)
$$\begin{array}{r} 128 \\ \times\ 46 \\ \hline \end{array}$$

(9)
$$\begin{array}{r} 262 \\ \times\ 23 \\ \hline \end{array}$$

(10)
$$\begin{array}{r} 785 \\ \times\ 28 \\ \hline \end{array}$$

(11)
$$\begin{array}{r} 581 \\ \times\ 35 \\ \hline \end{array}$$

(12)
$$\begin{array}{r} 421 \\ \times\ 45 \\ \hline \end{array}$$

Name ______________________ Date ______________

Multiply.

(1) 258×96

(2) 129×63

(3) 287×86

(4) 871×59

(5) 495×93

(6) 678×26

(7) 801×52

(8) 824×87

(9) 135×57

(10) 526×35

(11) 968×37

(12) 897×82

Name ______________________ Date ______________________

Multiply.

① $\begin{array}{r} 293 \\ \times\ 22 \\ \hline \end{array}$

② $\begin{array}{r} 124 \\ \times\ 82 \\ \hline \end{array}$

③ $\begin{array}{r} 863 \\ \times\ 34 \\ \hline \end{array}$

④ $\begin{array}{r} 407 \\ \times\ 29 \\ \hline \end{array}$

⑤ $\begin{array}{r} 454 \\ \times\ 42 \\ \hline \end{array}$

⑥ $\begin{array}{r} 863 \\ \times\ 56 \\ \hline \end{array}$

⑦ $\begin{array}{r} 105 \\ \times\ 46 \\ \hline \end{array}$

⑧ $\begin{array}{r} 394 \\ \times\ 26 \\ \hline \end{array}$

⑨ $\begin{array}{r} 437 \\ \times\ 28 \\ \hline \end{array}$

⑩ $\begin{array}{r} 971 \\ \times\ 49 \\ \hline \end{array}$

⑪ $\begin{array}{r} 261 \\ \times\ 58 \\ \hline \end{array}$

⑫ $\begin{array}{r} 813 \\ \times\ 43 \\ \hline \end{array}$

Name ____________________ Date ____________________

Multiply.

① 168 × 23

② 387 × 46

③ 252 × 64

④ 826 × 49

⑤ 791 × 23

⑥ 475 × 22

⑦ 795 × 92

⑧ 257 × 75

⑨ 323 × 67

⑩ 645 × 29

⑪ 734 × 43

⑫ 326 × 84

Name ______________________ Date ______________

Multiply.

① 876 × 42	② 628 × 25	③ 304 × 43
④ 167 × 39	⑤ 321 × 23	⑥ 723 × 96
⑦ 472 × 47	⑧ 847 × 59	⑨ 924 × 67
⑩ 864 × 36	⑪ 324 × 93	⑫ 371 × 35

Name ______________________________ Date ____________________

Multiply.

① $\begin{array}{r} 491 \\ \times\ 67 \\ \hline \end{array}$

② $\begin{array}{r} 185 \\ \times\ 18 \\ \hline \end{array}$

③ $\begin{array}{r} 969 \\ \times\ 49 \\ \hline \end{array}$

④ $\begin{array}{r} 587 \\ \times\ 75 \\ \hline \end{array}$

⑤ $\begin{array}{r} 525 \\ \times\ 93 \\ \hline \end{array}$

⑥ $\begin{array}{r} 345 \\ \times\ 87 \\ \hline \end{array}$

⑦ $\begin{array}{r} 307 \\ \times\ 96 \\ \hline \end{array}$

⑧ $\begin{array}{r} 354 \\ \times\ 75 \\ \hline \end{array}$

⑨ $\begin{array}{r} 835 \\ \times\ 38 \\ \hline \end{array}$

⑩ $\begin{array}{r} 326 \\ \times\ 49 \\ \hline \end{array}$

⑪ $\begin{array}{r} 429 \\ \times\ 23 \\ \hline \end{array}$

⑫ $\begin{array}{r} 287 \\ \times\ 58 \\ \hline \end{array}$

Name ____________________ Date ____________________

Multiply.

(1) 413×35

(2) 424×93

(3) 917×16

(4) 307×73

(5) 823×65

(6) 642×39

(7) 783×97

(8) 935×28

(9) 753×23

(10) 543×85

(11) 261×32

(12) 234×53

Name ______________________ Date ______________

Multiply.

① 362 × 52

② 347 × 47

③ 423 × 73

④ 158 × 68

⑤ 675 × 28

⑥ 872 × 53

⑦ 732 × 26

⑧ 565 × 24

⑨ 659 × 67

⑩ 902 × 63

⑪ 857 × 28

⑫ 205 × 19

Name ______________________ Date ______________

Multiply.

① 548×68

② 475×45

③ 184×26

④ 981×64

⑤ 528×97

⑥ 207×91

⑦ 601×72

⑧ 176×49

⑨ 314×75

⑩ 392×34

⑪ 678×83

⑫ 146×47

Name ______________________ Date ______________

Multiply.

(1) 595×53

(2) 492×37

(3) 475×82

(4) 496×34

(5) 646×49

(6) 364×54

(7) 649×36

(8) 968×52

(9) 932×52

(10) 301×74

(11) 185×24

(12) 477×31

Name ______________________ Date ______________________

Multiply.

① 476 × 13 =

② 828 × 17 =

③ 979 × 62 =

④ 681 × 84 =

⑤ 407 × 35 =

⑥ 328 × 55 =

⑦ 697 × 37 =

⑧ 345 × 48 =

⑨ 502 × 51 =

⑩ 523 × 45 =

⑪ 845 × 82 =

⑫ 167 × 78 =

Name ______________________ Date ______________

Multiply.

1.
```
  957
×  58
-----
```

2.
```
  261
×  78
-----
```

3.
```
  274
×  14
-----
```

4.
```
  383
×  47
-----
```

5.
```
  624
×  79
-----
```

6.
```
  562
×  65
-----
```

7.
```
  235
×  49
-----
```

8.
```
  127
×  85
-----
```

9.
```
  454
×  82
-----
```

10.
```
  175
×  63
-----
```

11.
```
  216
×  39
-----
```

12.
```
  246
×  18
-----
```

Name ______________________ Date ______________

Multiply.

(1) $\begin{array}{r} 434 \\ \times\ 72 \\ \hline \end{array}$

(2) $\begin{array}{r} 324 \\ \times\ 85 \\ \hline \end{array}$

(3) $\begin{array}{r} 823 \\ \times\ 23 \\ \hline \end{array}$

(4) $\begin{array}{r} 371 \\ \times\ 74 \\ \hline \end{array}$

(5) $\begin{array}{r} 987 \\ \times\ 46 \\ \hline \end{array}$

(6) $\begin{array}{r} 535 \\ \times\ 37 \\ \hline \end{array}$

(7) $\begin{array}{r} 282 \\ \times\ 85 \\ \hline \end{array}$

(8) $\begin{array}{r} 904 \\ \times\ 72 \\ \hline \end{array}$

(9) $\begin{array}{r} 451 \\ \times\ 53 \\ \hline \end{array}$

(10) $\begin{array}{r} 467 \\ \times\ 62 \\ \hline \end{array}$

(11) $\begin{array}{r} 826 \\ \times\ 92 \\ \hline \end{array}$

(12) $\begin{array}{r} 525 \\ \times\ 49 \\ \hline \end{array}$

Name ______________________ Date ______________

Multiply.

(1) $\begin{array}{r} 452 \\ \times\ 29 \\ \hline \end{array}$

(2) $\begin{array}{r} 281 \\ \times\ 34 \\ \hline \end{array}$

(3) $\begin{array}{r} 479 \\ \times\ 74 \\ \hline \end{array}$

(4) $\begin{array}{r} 352 \\ \times\ 62 \\ \hline \end{array}$

(5) $\begin{array}{r} 705 \\ \times\ 31 \\ \hline \end{array}$

(6) $\begin{array}{r} 843 \\ \times\ 23 \\ \hline \end{array}$

(7) $\begin{array}{r} 385 \\ \times\ 44 \\ \hline \end{array}$

(8) $\begin{array}{r} 796 \\ \times\ 68 \\ \hline \end{array}$

(9) $\begin{array}{r} 572 \\ \times\ 96 \\ \hline \end{array}$

(10) $\begin{array}{r} 272 \\ \times\ 42 \\ \hline \end{array}$

(11) $\begin{array}{r} 102 \\ \times\ 23 \\ \hline \end{array}$

(12) $\begin{array}{r} 659 \\ \times\ 53 \\ \hline \end{array}$

Name ____________________ Date ____________________

Multiply.

1. 578×45

2. 323×42

3. 218×95

4. 254×29

5. 386×68

6. 743×53

7. 674×68

8. 405×15

9. 594×79

10. 891×47

11. 597×43

12. 146×63

Name ______________________ Date ______________

Multiply.

① 757 × 49

② 154 × 75

③ 857 × 58

④ 706 × 24

⑤ 343 × 59

⑥ 585 × 42

⑦ 195 × 35

⑧ 783 × 94

⑨ 728 × 35

⑩ 298 × 74

⑪ 592 × 64

⑫ 928 × 39

Name ______________________ Date ______________

Multiply.

① $\begin{array}{r} 459 \\ \times\ 83 \\ \hline \end{array}$

② $\begin{array}{r} 536 \\ \times\ 35 \\ \hline \end{array}$

③ $\begin{array}{r} 384 \\ \times\ 72 \\ \hline \end{array}$

④ $\begin{array}{r} 538 \\ \times\ 42 \\ \hline \end{array}$

⑤ $\begin{array}{r} 802 \\ \times\ 76 \\ \hline \end{array}$

⑥ $\begin{array}{r} 534 \\ \times\ 25 \\ \hline \end{array}$

⑦ $\begin{array}{r} 135 \\ \times\ 92 \\ \hline \end{array}$

⑧ $\begin{array}{r} 567 \\ \times\ 53 \\ \hline \end{array}$

⑨ $\begin{array}{r} 269 \\ \times\ 68 \\ \hline \end{array}$

⑩ $\begin{array}{r} 475 \\ \times\ 51 \\ \hline \end{array}$

⑪ $\begin{array}{r} 627 \\ \times\ 38 \\ \hline \end{array}$

⑫ $\begin{array}{r} 846 \\ \times\ 54 \\ \hline \end{array}$

Name ______________________ Date ______________

Multiply.

① 362 × 35 =

② 583 × 59 =

③ 246 × 25 =

④ 462 × 94 =

⑤ 961 × 68 =

⑥ 825 × 59 =

⑦ 727 × 62 =

⑧ 862 × 41 =

⑨ 686 × 84 =

⑩ 903 × 46 =

⑪ 248 × 77 =

⑫ 565 × 28 =

Name ______________________ Date ______________

Multiply.

① 537 × 43 =

② 757 × 79 =

③ 515 × 36 =

④ 906 × 38 =

⑤ 741 × 24 =

⑥ 863 × 43 =

⑦ 518 × 39 =

⑧ 527 × 28 =

⑨ 326 × 65 =

⑩ 674 × 73 =

⑪ 531 × 54 =

⑫ 748 × 25 =

Name ____________________ Date ____________________

Multiply.

(1) 387 × 39 =

(2) 265 × 53 =

(3) 475 × 65 =

(4) 134 × 28 =

(5) 635 × 58 =

(6) 795 × 39 =

(7) 651 × 87 =

(8) 157 × 14 =

(9) 538 × 93 =

(10) 892 × 92 =

(11) 562 × 53 =

(12) 408 × 94 =

Name ______________________ Date ______________

Multiply.

(1) $\begin{array}{r} 484 \\ \times\ 38 \\ \hline \end{array}$

(2) $\begin{array}{r} 565 \\ \times\ 93 \\ \hline \end{array}$

(3) $\begin{array}{r} 474 \\ \times\ 43 \\ \hline \end{array}$

(4) $\begin{array}{r} 435 \\ \times\ 67 \\ \hline \end{array}$

(5) $\begin{array}{r} 818 \\ \times\ 67 \\ \hline \end{array}$

(6) $\begin{array}{r} 501 \\ \times\ 74 \\ \hline \end{array}$

(7) $\begin{array}{r} 259 \\ \times\ 74 \\ \hline \end{array}$

(8) $\begin{array}{r} 464 \\ \times\ 17 \\ \hline \end{array}$

(9) $\begin{array}{r} 628 \\ \times\ 35 \\ \hline \end{array}$

(10) $\begin{array}{r} 834 \\ \times\ 87 \\ \hline \end{array}$

(11) $\begin{array}{r} 314 \\ \times\ 42 \\ \hline \end{array}$

(12) $\begin{array}{r} 579 \\ \times\ 47 \\ \hline \end{array}$

SECTION

5

Multiplication (Challenging)

3-Digits x 3-Digits

		2 1		
	3	0	7	
x	1	4	2	
		6	1	4
1	2	2	8	0
3	0	7	0	0
4	3,	5	9	4

STEP 1

Multiply by the ones digit of 142.

Multiply: 2 x 307 = 614

STEP 2

Multiply by the tens digit of 142. Write the product on the next row with 0 in the ones column (since 4 is the tens digit).

Write 0 in the ones column on the next row.
Multiply: 4 x 307= 1,228 *(We have 12,280 since 0 is in the ones place)*

STEP 3

Multiply by the hundreds digit of 142. Write the product on the next row row with 0 in the ones column and the tens column (since 1 is the hundreds digit).

Write 0 in the ones column and the tens column on the next row.
Multiply: 1 x 307 = 307 *(We have 30,700 since 0's are in the ones and tens places.)*

STEP 4

Add the three rows.

Add: 614 + 12,280 + 30,700 = 43,594

Name ____________________ Date ____________________

Multiply.

(1)
$$\begin{array}{r} 214 \\ \times\ 296 \\ \hline \end{array}$$

(2)
$$\begin{array}{r} 963 \\ \times\ 503 \\ \hline \end{array}$$

(3)
$$\begin{array}{r} 764 \\ \times\ 934 \\ \hline \end{array}$$

(4)
$$\begin{array}{r} 293 \\ \times\ 482 \\ \hline \end{array}$$

(5)
$$\begin{array}{r} 347 \\ \times\ 258 \\ \hline \end{array}$$

(6)
$$\begin{array}{r} 475 \\ \times\ 417 \\ \hline \end{array}$$

(7)
$$\begin{array}{r} 654 \\ \times\ 424 \\ \hline \end{array}$$

(8)
$$\begin{array}{r} 463 \\ \times\ 741 \\ \hline \end{array}$$

(9)
$$\begin{array}{r} 325 \\ \times\ 564 \\ \hline \end{array}$$

Name ______________________ Date ______________

Multiply.

(1)
$$\begin{array}{r} 734 \\ \times\ 321 \\ \hline \end{array}$$

(2)
$$\begin{array}{r} 231 \\ \times\ 985 \\ \hline \end{array}$$

(3)
$$\begin{array}{r} 647 \\ \times\ 209 \\ \hline \end{array}$$

(4)
$$\begin{array}{r} 732 \\ \times\ 868 \\ \hline \end{array}$$

(5)
$$\begin{array}{r} 853 \\ \times\ 271 \\ \hline \end{array}$$

(6)
$$\begin{array}{r} 758 \\ \times\ 265 \\ \hline \end{array}$$

(7)
$$\begin{array}{r} 204 \\ \times\ 402 \\ \hline \end{array}$$

(8)
$$\begin{array}{r} 649 \\ \times\ 248 \\ \hline \end{array}$$

(9)
$$\begin{array}{r} 152 \\ \times\ 579 \\ \hline \end{array}$$

Name ______________________ Date ______________

Multiply.

(1)
$$\begin{array}{r} 927 \\ \times\ 529 \\ \hline \end{array}$$

(2)
$$\begin{array}{r} 723 \\ \times\ 548 \\ \hline \end{array}$$

(3)
$$\begin{array}{r} 284 \\ \times\ 862 \\ \hline \end{array}$$

(4)
$$\begin{array}{r} 483 \\ \times\ 286 \\ \hline \end{array}$$

(5)
$$\begin{array}{r} 857 \\ \times\ 604 \\ \hline \end{array}$$

(6)
$$\begin{array}{r} 861 \\ \times\ 846 \\ \hline \end{array}$$

(7)
$$\begin{array}{r} 524 \\ \times\ 769 \\ \hline \end{array}$$

(8)
$$\begin{array}{r} 187 \\ \times\ 816 \\ \hline \end{array}$$

(9)
$$\begin{array}{r} 504 \\ \times\ 152 \\ \hline \end{array}$$

Name ______________________ Date ______________

Multiply.

1)
```
  8 5 1
× 7 5 7
```

2)
```
  2 4 5
× 3 2 5
```

3)
```
  2 7 9
× 6 4 7
```

4)
```
  9 7 4
× 6 2 3
```

5)
```
  2 0 2
× 1 4 8
```

6)
```
  1 5 7
× 3 8 3
```

7)
```
  4 8 2
× 6 7 8
```

8)
```
  1 8 1
× 8 1 5
```

9)
```
  3 2 6
× 5 3 8
```

Name ______________________ Date ______________________

Multiply.

1)
```
   2 7 8
 × 3 2 6
 -------
```

2)
```
   5 3 5
 × 2 4 4
 -------
```

3)
```
   8 5 4
 × 4 9 5
 -------
```

4)
```
   4 1 2
 × 9 8 4
 -------
```

5)
```
   8 6 7
 × 9 2 8
 -------
```

6)
```
   8 7 8
 × 6 2 5
 -------
```

7)
```
   3 7 8
 × 2 2 0
 -------
```

8)
```
   6 7 4
 × 1 8 3
 -------
```

9)
```
   5 6 7
 × 5 2 5
 -------
```

SECTION

A

Answer Key

Answers to problems.

Answer Key

PAGE 2

128	273	115	651	108
246	162	88	54	332
295	196	459	144	98
343	140	135	189	483

PAGE 3

129	250	58	498	270
296	322	243	188	480
102	144	486	340	402
448	291	522	70	148

PAGE 4

175	354	426	184	448
258	290	112	210	104
62	232	110	558	220
48	416	644	455	399

PAGE 5

118	180	28	236	66
432	415	100	305	174
128	312	680	245	156
423	54	553	76	198

PAGE 6

156	234	792	342	84
228	150	312	736	210
174	472	102	189	468
325	304	148	440	171

PAGE 7

72	410	512	138	125
355	256	368	558	328
164	170	243	504	378
84	531	285	144	96

PAGE 8

496	360	828	168	204
632	372	336	216	300
435	126	69	282	238
684	174	380	104	240

PAGE 9

200	588	150	182	360
441	91	190	192	420
384	600	190	873	352
273	336	74	284	261

PAGE 10

468	567	301	264	216
106	42	657	96	26
444	760	80	612	486
145	475	315	168	146

PAGE 11

456	693	232	50	375
144	231	588	744	165
288	356	152	132	110
154	144	136	105	162

PAGE 12

120	142	160	117	52
196	46	465	177	536
696	252	592	504	38
552	105	335	72	609

PAGE 13

248	114	385	68	120
162	576	511	364	477
532	90	124	264	134
175	60	756	376	324

PAGE 14

260	186	220	96	675
119	95	172	225	130
406	168	340	315	153
72	111	75	204	320

PAGE 15

765	328	51	438	188
265	370	664	70	420
330	222	261	288	267
316	135	324	252	166

PAGE 16

225	285	56	344	396
658	608	133	93	195
217	207	582	616	396
170	86	78	455	185

PAGE 17

112	64	672	574	280
488	172	368	637	430
159	315	470	584	528
855	205	225	648	87

PAGE 18

184	495	196	85	665
492	224	248	32	525
595	180	282	752	729
440	294	63	656	270

PAGE 19

116	186	276	801	310
30	215	324	228	712
210	864	122	108	357
688	98	48	182	747

PAGE 20

224	36	192	198	92
837	150	102	252	490
162	81	405	135	434
245	249	518	192	126

PAGE 21

450	648	140	427	230
136	207	544	104	388
155	623	68	45	432
296	378	522	308	57

PAGE 22

774	348	464	513	260
42	424	234	255	768
385	178	376	392	116
105	255	686	78	365

PAGE 24

1,460	7,749	2,604	3,832
1,890	4,950	3,635	1,874
2,370	1,752	2,310	1,636
3,792	3,608	1,224	1,896

PAGE 25

1,645	1,272	4,025	2,732
5,868	2,886	1,038	3,332
4,620	1,974	1,518	1,808
3,836	2,260	1,690	4,305

PAGE 26

1,950	4,711	4,608	6,048
2,484	1,962	3,144	3,160
1,095	1,288	2,630	2,200
516	1,377	2,984	3,606

PAGE 27

3,492	3,375	1,652	3,516
732	1,904	2,160	2,346
1,516	1,278	1,484	4,067
4,865	2,492	2,073	7,803

PAGE 28

2,853	5,792	1,024	815
6,622	5,832	4,232	2,710
2,037	3,312	2,128	1,020
3,150	657	3,038	830

PAGE 29

2,889	3,115	3,712	945
3,640	2,748	2,002	1,173
1,263	1,203	1,626	952
5,610	1,080	5,201	1,704

PAGE 30

2,184	2,032	5,364	4,428
1,476	6,517	2,936	5,691
835	4,365	1,538	546
6,776	4,168	2,315	966

PAGE 31

940	1,995	648	6,608
2,112	2,332	6,678	316
2,205	6,344	1,872	4,984
1,396	1,611	2,124	652

PAGE 32

1,922	2,376	660	4,878
5,957	2,316	1,668	4,090
1,944	2,548	2,534	3,630
5,859	4,344	828	1,188

PAGE 33

1,542	1,374	1,712	801
496	7,000	4,446	1,848
4,445	1,455	1,522	4,592
978	4,338	2,605	6,076

PAGE 34

1,056	1,296	822	1,210
3,232	2,868	3,684	3,810
3,126	1,395	4,018	4,776
2,172	1,920	630	3,048

PAGE 35

25,208	7,524	13,498
33,341	67,496	22,596
43,255	60,633	35,413
31,456	15,634	12,744

PAGE 36

53,800	34,569	36,365
34,240	58,092	7,772
25,767	7,566	14,112
30,268	29,696	16,716

PAGE 37

53,694	8,897	18,172
26,160	16,152	27,216
64,408	7,136	51,512
4,638	59,656	36,240

PAGE 38

29,694	14,068	42,687
32,865	18,455	44,448
69,125	25,746	19,674
17,961	67,896	19,467

PAGE 39

27,684	47,544	17,043
60,208	23,715	39,180
57,069	26,592	51,642
3,342	16,251	27,726

PAGE 40

29,772	16,425	45,135
43,137	13,860	32,438
10,504	29,478	6,546
17,872	14,542	43,491

PAGE 41

15,270	30,524	5,488
64,424	13,892	9,162
67,081	10,686	25,092
17,236	75,048	18,948

PAGE 42

37,905	47,824	11,884
19,755	11,848	36,218
50,715	14,076	18,275
25,710	37,945	16,284

PAGE 43

14,470	19,676	16,374
4,074	60,578	63,756
5,618	17,472	23,810
40,736	26,142	3,930

PAGE 44

25,060	41,118	15,831
13,038	57,295	56,778
46,432	28,840	14,780
41,502	82,512	11,752

PAGE 46

5,472	2,160	304	3,145
884	2,835	972	816
3,417	1,344	2,925	2,773
2,944	2,822	2,775	576

PAGE 47

2,914	2,809	3,913	2,072
4,588	2,173	1,656	5,952
4,779	1,368	3,977	3,610
5,922	351	2,958	3,735

PAGE 48

1,872	728	4,526	3,219
702	3,174	4,335	4,272
4,277	2,394	1,120	2,304
1,820	725	5,832	1,872

PAGE 49

621	6,675	1,104	1,612
1,512	2,047	1,638	2,848
1,608	3,071	806	1,833
5,561	2,765	2,109	7,663

PAGE 50

2,523	3,068	1,725	6,916
1,118	4,272	6,768	1,479
988	1,300	6,156	1,404
950	2,795	325	2,968

PAGE 51

1,568	4,324	2,890	4,508
1,944	1,230	2,419	4,992
1,938	3,555	1,632	3,219
1,350	3,770	1,647	1,975

PAGE 52

4,698	768	2,065	806
1,911	4,838	1,875	4,030
2,226	224	3,182	2,350
810	1,197	2,408	5,766

PAGE 53

546	4,845	1,225	4,324
6,164	3,772	1,739	768
6,417	3,984	4,176	900
480	1,554	6,450	4,464

PAGE 54

3,948	2,438	6,225	7,553
3,948	1,377	799	2,001
4,736	6,045	720	1,702
2,982	5,418	3,182	3,384

PAGE 55

4,758	1,849	2,414	5,369
1,288	2,268	1,295	2,325
1,275	5,644	6,474	2,496
1,216	2,520	1,633	3,220

PAGE 56

1,984	1,890	2,077	4,845
391	2,349	7,544	2,688
2,842	1,044	5,041	2,924
6,486	2,622	756	1,792

PAGE 57

2,652	2,135	1,932	1,829
5,022	3,024	588	5,256
1,961	4,536	1,944	2,550
2,914	8,051	966	4,371

PAGE 58

322	7,885	3,948	6,120
3,854	4,929	4,582	375
403	3,354	7,104	1,323
360	3,220	3,182	3,478

PAGE 59

1,794	1,886	1,890	4,374
2,183	2,576	5,890	2,142
6,351	3,422	1,484	3,358
1,533	2,752	3,239	2,392

PAGE 60

3,276	4,320	2,175	196
805	7,917	7,347	943
3,024	1,125	4,902	3,024
1,148	3,813	2,208	5,320

PAGE 61

2,652	1,080	3,774	6,586
5,846	1,560	1,794	1,495
1,134	1,700	1,092	1,445
3,283	1,534	6,640	4,650

PAGE 62

3,886	1,394	1,075	731
1,248	5,185	2,967	7,905
3,082	4,183	299	6,351
2,907	3,264	1,134	612

PAGE 63

1,530	690	5,518	3,468
1,428	1,600	1,204	3,848
3,471	2,275	3,552	1,701
1,692	3,042	1,728	2,625

PAGE 64

5,256	2,170	490	1,215
1,800	4,512	2,100	3,496
5,084	5,487	1,054	7,644
1,245	1,520	2,646	4,964

PAGE 65

2,394	1,344	3,216	3,132
1,540	3,075	5,074	2,444
602	2,806	1,537	1,702
266	5,432	2,128	2,380

PAGE 66

2,912	3,192	6,566	1,537
3,496	600	1,577	7,626
4,278	888	645	1,085
1,395	3,354	2,808	1,300

PAGE 68

6,494	23,643	17,052
16,468	27,892	86,296
39,825	5,888	6,026
21,980	20,335	18,945

PAGE 69

24,768	8,127	24,682
51,389	46,035	17,628
41,652	71,688	7,695
18,410	35,816	73,554

PAGE 70

6,446	10,168	29,342
11,803	19,068	48,328
4,830	10,244	12,236
47,579	15,138	34,959

PAGE 71

3,864	17,802	16,128
40,474	18,193	10,450
73,140	19,275	21,641
18,705	31,562	27,384

PAGE 72

36,792	15,700	13,072
6,513	7,383	69,408
22,184	49,973	61,908
31,104	30,132	12,985

PAGE 73

32,897	3,330	47,481
44,025	48,825	30,015
29,472	26,550	31,730
15,974	9,867	16,646

PAGE 74

14,455	39,432	14,672
22,411	53,495	25,038
75,951	26,180	17,319
46,155	8,352	12,402

PAGE 75

18,824	16,309	30,879
10,744	18,900	46,216
19,032	13,560	44,153
56,826	23,996	3,895

PAGE 76

37,264	21,375	4,784
62,784	51,216	18,837
43,272	8,624	23,550
13,328	56,274	6,862

PAGE 77

31,535	18,204	38,950
16,864	31,654	19,656
23,364	50,336	48,464
22,274	4,440	14,787

PAGE 78

6,188	14,076	60,698
57,204	14,245	18,040
25,789	16,560	25,602
23,535	69,290	13,026

PAGE 79

55,506	20,358	3,836
18,001	49,296	36,530
11,515	10,795	37,228
11,025	8,424	4,428

PAGE 80

31,248	27,540	18,929
27,454	45,402	19,795
23,970	65,088	23,903
28,954	75,992	25,725

PAGE 81

13,108	9,554	35,446
21,824	21,855	19,389
16,940	54,128	54,912
11,424	2,346	34,927

PAGE 82

26,010	13,566	20,710
7,366	26,248	39,379
45,832	6,075	46,926
41,877	25,671	9,198

PAGE 83

37,093	11,550	49,706
16,944	20,237	24,570
6,825	73,602	25,480
22,052	37,888	36,192

PAGE 84

38,097	18,760	27,648
22,596	60,952	13,350
12,420	30,051	18,292
24,225	23,826	45,684

PAGE 85

12,670	34,397	6,150
43,428	65,348	48,675
45,074	35,342	57,624
41,538	19,096	15,820

PAGE 86

23,091	59,803	18,540
34,428	17,784	37,109
20,202	14,756	21,190
49,202	28,674	18,700

PAGE 87

15,093	14,045	30,875
3,752	36,830	31,005
56,637	2,198	50,034
82,064	29,786	38,352

PAGE 88

18,392	52,545	20,382
29,145	54,806	37,074
19,166	7,888	21,980
72,558	13,188	27,213

PAGE 90

63,344	484,389	713,576
141,226	89,526	198,075
277,296	343,083	183,300

PAGE 91

235,614	227,535	135,223
635,376	231,163	200,870
82,008	160,952	88,008

PAGE 92

490,383	396,204	244,808
138,138	517,628	728,406
402,956	152,592	76,608

PAGE 93

644,207	79,625	180,513
606,802	29,896	60,131
326,796	147,515	175,388

PAGE 94

90,628	130,540	422,730
405,408	804,576	548,750
83,160	123,342	297,675

We hope you found our book helpful.
Please consider leaving a review.